EARLY BIRD STORIES™

Let's Notice Animal Behavior

Martha E. H. Rustad   Illustrated by **Mike Moran**

LERNER PUBLICATIONS ◆ **MINNEAPOLIS**

NOTE TO EDUCATORS

Find text recall questions at the end of each chapter. Critical-thinking and text feature questions are available on page 23. These help young readers learn to think critically about the topic by using the text, text features, and illustrations.

Lerner Publications Company
An imprint of Lerner Publishing Group, Inc.
241 First Avenue North
Minneapolis, MN 55401 USA

For reading levels and more information, look up this title at www.lernerbooks.com.

Photos on p. 22 used with permission of: Acon Cheng/Shutterstock.com (monkeys); Johnathan Esper/Shutterstock.com (elephant herd); Bildagentur Zoonar GmbH/Shutterstock.com (tiger cub).

Main body text set in Billy Infant.
Typeface provided by SparkyType.

Library of Congress Cataloging-in-Publication Data

Names: Rustad, Martha E. H. (Martha Elizabeth Hillman), 1975- author. | Moran, Mike, illustrator.
Title: Let's notice animal behavior / Martha E.H. Rustad; illustrated by Mike Moran.
Description: Minneapolis : Lerner Publications, [2022] | Series: Let's make observations (early bird stories) | Includes bibliographical references and index. | Audience: Ages 5-8 | Audience: Grades K-1 | Summary: "Birds chirp, river otters swim, lions pounce, and capuchin monkeys groom. Readers will delight in learning alongside Ms. McLean's class about animals' behavior in this charming, illustrated story"—Provided by publisher.
Identifiers: LCCN 2021010469 (print) | LCCN 2021010470 (ebook) | ISBN 9781728441337 (library binding) | ISBN 9781728444642 (ebook)
Subjects: LCSH: Animal behavior—Juvenile literature.
Classification: LCC QL751.5 .R885 2022 (print) | LCC QL751.5 (ebook) | DDC 591.5—dc23

LC record available at https://lccn.loc.gov/2021010469
LC ebook record available at https://lccn.loc.gov/2021010470

Manufactured in the United States of America

TABLE OF CONTENTS

CHAPTER 1

FIELD TRIP!

Our class is going to the zoo! We will **observe** the animals' **behaviors** and record them in charts.

"Behavior is the way an animal acts," says our teacher, Ms. McLean. "Observe is to watch something closely."

ANIMAL				
Making Noise				
Eating				
Playing				
Walking				
Flying				
Swimming				
Other				

✓Check!
What do *behavior* and *observe* mean?

5

CHAPTER 2

ZOO OBSERVATIONS

Mr. Sato's a zookeeper. He takes care of animals and teaches zoo visitors. He shows us different animals.

A tiger cub **meows**. The mother tiger feeds it.

When a zebra foal **whinnies**, its mother feeds it.

"Different types of animals sometimes act the same," Mr. Sato says.

	TIGER	ZEBRA
Making Noise	X	X
Eating	X	X
Walking	X	X

We have a snack break under a tree.

A robin feeds worms to its chirping chicks.

✓Check! How is the tiger cub's behavior and the zebra foal's behavior alike?

CHAPTER 3

ANIMAL LEARNING

River otters **teach** their pups to swim and dive.

Raccoons teach their kits how to find food. The kits get hungry from running and playing!

	OTTER	RACCOON
Eating		X
Playing		X
Swimming	X	

Lion cubs play, pounce, and practice hunting.

Young chimpanzees **learn** to use tools from adult chimps. They use long sticks to catch ants.

"Many young animals learn from adult animals," Mr. Sato says.

"Just like kids learn from teachers at school," Ms. McLean adds.

✓Check! What do young animals learn from their parents?

15

CHAPTER 4

HELPING THE GROUP

Elephants live in family groups.

If a predator comes, adult elephants circle the young to **protect** them.

	LION	CHIMP	ELEPHANT
			X
Eating			
Playing	X		
Walking			X
Sitting		X	
Other		X	

Capuchin monkeys **groom**. They pick bugs off one another's fur and eat them.

"Elephants and monkeys both live in groups!" says Raj.

"And they all help take care of others," Tonia adds.

Our zoo trip is over.
Together we say, "Thank you, Mr. Sato!"

Back at our classroom, we talk about all
of the animal behaviors we saw today.

✓Check! What is
one behavior elephants
and monkeys have in
common?

LEARN ABOUT ANIMAL BEHAVIOR

Like many baby animals, zebra foals drink their mother's milk until they are old enough to eat solid food.

While some animals live alone, many animals live in groups.

Animals take care of one another in different ways. Monkeys groom. They grab bugs off another monkey's fur to help clean them.

Baby animals can learn skills at a young age. Tiger cubs learn to pounce and hunt before they turn one.

Most animals have predators. When a predator approaches an elephant herd, the adult elephants surround their young to protect them.

THINK ABOUT ANIMAL BEHAVIOR:
CRITICAL-THINKING AND TEXT FEATURE QUESTIONS

How does your favorite animal behave?

What animals can you see near your home?

Can you find the glossary?

Who is the author of this book?

Expand learning beyond the printed book. Download free, complementary educational resources for this book from our website, www.lerneresource.com.

GLOSSARY

behavior: the way a person or animal acts

groom: to clean an animal's fur

observe: to watch something closely

pounce: to leap or jump suddenly

predator: an animal that hunts and eats other animals

LEARN MORE

Andrus, Aubre. *A Zebra's Day.* Washington, DC: National Geographic Kids, 2020.

Clark, Rosalyn. *A Visit to the Zoo.* Minneapolis: Lerner Publications, 2018.

***National Geographic Kids*—Zoo Animal Field Trips**
https://kids.nationalgeographic.com/animals/article/zoo-animal-field-trips

INDEX